ENGINEERING FOR DISASTER

ENGINEERING FOR HURRICANES

by Wendy Hinote Lanier

FOCUS READERS.
NAVIGATOR

WWW.FOCUSREADERS.COM

Copyright © 2021 by Focus Readers®, Lake Elmo, MN 55042. All rights reserved. No part of this book may be reproduced or utilized in any form or by any means without written permission from the publisher.

Focus Readers is distributed by North Star Editions:
sales@northstareditions.com | 888-417-0195

Produced for Focus Readers by Red Line Editorial.

Content Consultant: David Roueche, Assistant Professor of Civil Engineering, Auburn University

Photographs ©: Shutterstock Images, cover, 1, 8–9, 13, 15, 16–17, 19, 21; iStockphoto, 4–5; Red Line Editorial, 7, 11; Wilfredo Lee/AP Images, 22–23; Andy Newman/AP Images, 25; Jim West/Alamy, 26–27; Gerald Herbert/AP Images, 29

Library of Congress Cataloging-in-Publication Data
Names: Lanier, Wendy Hinote, author.
Title: Engineering for hurricanes / by Wendy Hinote Lanier.
Description: Lake Elmo, MN : Focus Readers, 2021. | Series: Engineering for disaster | Includes index. | Audience: Grades 4–6
Identifiers: LCCN 2020002180 (print) | LCCN 2020002181 (ebook) | ISBN 9781644933800 (hardcover) | ISBN 9781644934562 (paperback) | ISBN 9781644936085 (ebook pdf) | ISBN 9781644935323 (hosted ebook)
Subjects: LCSH: Building, Stormproof--Juvenile literature. | Hurricane protection--Juvenile literature. | Flood damage prevention--Juvenile literature. | Levees--Design and construction--Juvenile literature.
Classification: LCC TH1096 .L36 2021 (print) | LCC TH1096 (ebook) | DDC 693.8/5--dc23
LC record available at https://lccn.loc.gov/2020002180
LC ebook record available at https://lccn.loc.gov/2020002181

Printed in the United States of America
Mankato, MN
082020

ABOUT THE AUTHOR
Wendy Hinote Lanier is a former elementary teacher and a Texan through and through. She writes and speaks for children and adults on a variety of topics. She is the author of more than 40 books for children and young adults on topics related to science, technology, social studies, the arts, and, of course, Texas.

TABLE OF CONTENTS

CHAPTER 1
Levee Breach 5

CHAPTER 2
Building Stormproof Communities 9

CASE STUDY
Water Management 14

CHAPTER 3
Building Strong Homes 17

CHAPTER 4
Weathering the Storm 23

CHAPTER 5
What's Next? 27

Focus on Engineering for Hurricanes • 30
Glossary • 31
To Learn More • 32
Index • 32

CHAPTER 1

LEVEE BREACH

On August 29, 2005, Hurricane Katrina blew ashore. The people of New Orleans, Louisiana, began to relax. The storm's strongest winds were to the east. Mississippi was getting hit badly. But New Orleans was protected by **levees**. Residents believed they were safe from the **storm surge**.

Hurricane Katrina's winds and floodwaters destroyed many communities.

Then, the unthinkable happened. The levees began to break. Soon, most of New Orleans was underwater. Nearly 400 people drowned in the rising water.

After the storm, engineers learned why the levees had failed. In some cases, the storm surge had risen above the levees. Water spilled over the top. It **eroded** the earth on the back side. Other parts of the levees failed for another reason. They had not been designed or built properly. During the storm, soil under the poorly built sections soaked up water. The **foundation** eroded. The levees fell.

Since Hurricane Katrina, workers have rebuilt some of the levees. But more work is needed to protect the city from future storms.

HOW THE LEVEES FAILED

The levees in New Orleans had two key parts. A steel sheet stretched 20 to 40 feet (6–12 m) into the ground. A concrete wall went on top of the steel sheet. The wall was 15 to 18 feet (4.6–5.5 m) high.

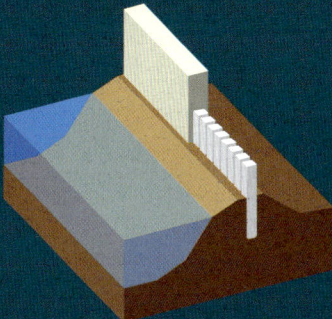 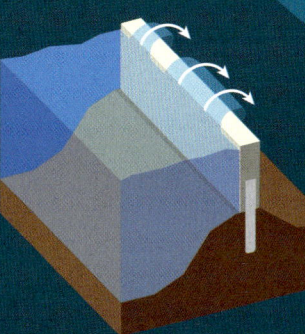

The water level usually wouldn't be high enough to reach the concrete walls.

Katrina's storm surge put pressure on the walls and ground. The storm surge was also high enough to pour over the walls.

The soil under the levees weakened. Water pushed the levees back, eventually causing the levees to collapse.

CHAPTER 2

BUILDING STORMPROOF COMMUNITIES

Hurricanes affect coastal areas. The best way to prevent hurricane damage is to not build in these areas. But many people want to live there anyway. So, engineers help these communities withstand the storms.

For example, bridges in hurricane zones face strong winds and high water.

Millions of people live on coasts, where they are at risk from hurricanes.

Engineers design bridges to withstand these forces. They make sure bridges are high enough to be above the storm surge. Sometimes they build openings in the bridge supports. The storm surge can pass through the openings. The water does not damage the bridge as much.

Engineers also design levees. The simplest levees are just mounds of soil. New Orleans' levees had concrete walls on top of metal sheets. The city had two types of walls. Some walls were shaped like the letter *I*. Other walls were shaped like an upside-down letter *T*.

After Hurricane Katrina, engineers found two problems. First, the I-walls did not stand up to the storm surge. The T-walls worked better. They could withstand some of the water's force. Second, the metal sheets under the levees did not go deep enough. They did not stop water from flowing underground.

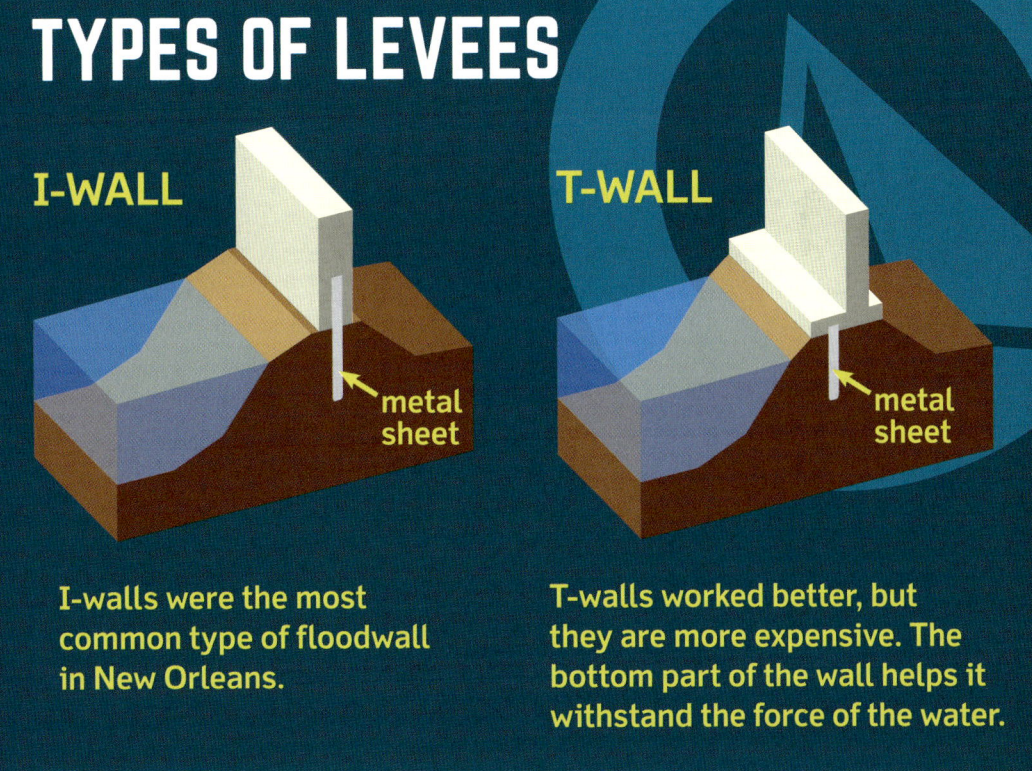

TYPES OF LEVEES

I-WALL — metal sheet

T-WALL — metal sheet

I-walls were the most common type of floodwall in New Orleans.

T-walls worked better, but they are more expensive. The bottom part of the wall helps it withstand the force of the water.

Now, engineers are using what they learned to build better levees.

Water control is an important part of preparing for hurricanes. Pipe systems carry water throughout a community. But too much water can overwork the systems. Hurricanes bring heavy rainfall

WHAT IS A HURRICANE?

Hurricanes are storms that form over warm tropical waters. They have winds of more than 74 miles per hour (119 km/h). The storms spin around a low-pressure center. Scientists rank hurricanes from 1 to 5. They base the rankings on wind speed. Hurricanes' high winds can be dangerous. But their storm surge and heavy rainfall cause the most damage.

Hurricane Harvey caused major flooding in Houston in 2017. Many people had to use boats to get around.

and storm surge. Engineers work to reduce this extra water. For example, they have created new paving materials. Usually water collects on top of the pavement. But new materials allow water to pass through to the soil underneath. The soil soaks up some of the extra water. That way, the city's pipe systems don't have to deal with as much water.

CASE STUDY

WATER MANAGEMENT

A city's pumping stations are important during big-rain events such as hurricanes. They pump water away from low-lying areas. They send the water to canals and other waterways. The pumps help prevent flooding. But overworked pumps could break down just when they are needed most.

After Hurricane Katrina, New Orleans changed the way it handled rain. City officials stopped allowing all rainwater to go into the pipe systems. Instead, they looked for ways to let water soak into the ground. They began using the landscape to store water. They created rain gardens and put out water barrels. They began to think of rain as a resource. And they developed plans to put it

In New Orleans, pumping stations are a key part of the city's flood-control efforts.

to better use. Their efforts meant the pumping stations are less likely to get overworked. Now, New Orleans is an example to other coastal cities.

CHAPTER 3

BUILDING STRONG HOMES

All cities have building codes that builders must follow. These rules make sure new buildings are safe and long-lasting. For example, across the United States, new buildings must be able to withstand winds of 95 miles per hour (150 km/h). In coastal areas at risk from hurricanes, this number is even higher.

Hurricanes can destroy homes that were not built well.

Building codes vary from city to city. But people can use common practices to help keep their homes safe from hurricanes.

Keeping wind out of a home is important. Wind inside a home increases the home's air pressure. The extra air pushes out against the walls. It presses up against the roof. High air pressure could cause the roof to pop off.

To keep wind from entering, homes can have **storm windows** or shutters. The shutters keep wind-blown objects from hitting and breaking windows.

The way parts of a house connect is also important. For example, workers can use nails rather than staples to hold down

Storm shutters protect this house's windows from flying debris.

shingles. They can connect the roof and the walls using **hurricane clips**. And they can use extra-long bolts to connect the walls and the foundation. These practices can help keep a house standing in a hurricane's strong winds.

In areas that flood often, engineers recommend building raised homes. A raised home stands on **stilts**. The stilts are made of concrete or another strong material. They raise the house more than

FLORIDA BUILDING CODES

In 1992, Hurricane Andrew destroyed entire communities near Miami, Florida. After the storm, state officials strengthened building codes in south Florida. In 2002, Florida created its first statewide building code designed to protect against wind. In 2018, Hurricane Michael hit an area of Florida that was not often affected by hurricanes. It heavily damaged several communities. Officials began to rethink how to apply the building codes. They planned to set tougher building codes statewide.

Concrete stilts protect a home against future floodwaters.

3 feet (1 m) above the area's average flood level.

Some homes have solid concrete foundations instead of stilts. Workers cut openings into the concrete. The openings let floodwater pass through the foundation. The water does less damage to the home.

CHAPTER 4

WEATHERING THE STORM

As a hurricane approaches, engineers get to work. Some engineers predict power outages. They study an area's land features, current weather, and power use. They create computer models based on their findings. Their models help power companies find areas likely to lose power.

A specialist monitors the movement of a hurricane as it approaches the coast.

Sometimes the models help companies prevent power outages.

Engineers also predict storm surge. They study local structures and land features. They figure out where the storm surge is likely to occur. They also predict how high it will be. Their work helps

THE 1900 GALVESTON STORM

In September 1900, a hurricane hit Galveston, Texas. It destroyed the city. Thousands of people died. After the storm, engineers raised the remaining buildings. They pumped sand underneath. This action raised the level of the entire city. Engineers also built a 17-foot (5.2-m) seawall to protect against storm surge. These two changes have helped Galveston weather many storms since.

Scientists track a hurricane's wind speed, location, and movement. They work with engineers to predict flooding.

officials put out timely flood warnings. It also helps officials plan search and rescue efforts. Officials know where people will most likely need help.

After the storm, engineers study data from **satellites**. They see how buildings and cities performed during the storm. They use this information to help cities prepare for the next big storm.

CHAPTER 5

WHAT'S NEXT?

Engineers have designed homes that can survive a hurricane's winds. Today's challenge is to make these homes less costly to build. Another challenge is keeping roof shingles and outer walls from letting water into homes during storms. Engineers continue to look for ways people can prepare for a storm.

Volunteers with Habitat for Humanity work to rebuild homes after Hurricane Katrina.

For example, engineers now tell people to close interior doors during hurricanes. If a window or door breaks, wind will rush into the home. But with closed doors, the increasing wind pressure doesn't spread through the home as easily. This practice reduces damage to the house. And it helps keep the roof from being blown off.

Many engineers are studying the way communities are built. In cities, land is often paved over. Engineers are working with community planners to change this. They hope to create landscapes that act as natural wetlands. These areas let water soak into the ground naturally. They reduce flooding.

Workers in Louisiana close a floodgate in preparation for Hurricane Nate in 2017.

Engineers are designing **floodgates** for areas that frequently flood. They look to the Netherlands as an example. Levees and floodgates have protected that country for more than 65 years.

Hurricanes are not new. But they are affecting more people as coastal populations grow. Engineers play a big role in helping these people weather the storms.

FOCUS ON
ENGINEERING FOR HURRICANES

Write your answers on a separate piece of paper.

1. Write a letter to a friend explaining what happened to the New Orleans levees during Hurricane Katrina.

2. Why do you think people continue to live in areas that are affected by hurricanes?

3. Which feature helps keep homes from flooding during hurricanes?
    - **A.** hurricane clips
    - **B.** storm windows
    - **C.** stilts

4. What might happen if engineers do not work to predict storm surge?
    - **A.** There would be no flood warnings.
    - **B.** There would be no flood.
    - **C.** Floodwaters would reach new heights.

Answer key on page 32.

GLOSSARY

eroded
Wore away slowly over time.

floodgates
Gates that can be opened or closed and help control the flow of water in low-lying areas.

foundation
The supporting base of a structure.

hurricane clips
Strong metal anchors that connect a roof to the walls and can resist high winds.

levees
Walls built from earth materials to stop floodwaters.

satellites
Objects or vehicles that orbit a planet or moon, often to collect information.

stilts
Poles or pillars that raise something off the ground.

storm surge
A rising of the sea that is caused by a storm's winds.

storm windows
Thick windows that provide extra protection against bad weather.

TO LEARN MORE

BOOKS

Gagne, Tammy. *Women in Engineering*. Minneapolis: Abdo Publishing, 2017.

Galat, Joan Marie. *Solve This! Wild and Wacky Challenges for the Genius Engineer in You*. Washington, DC: National Geographic Kids, 2018.

Smibert, Angie. *Environmental Engineering in the Real World*. Minneapolis: Abdo Publishing, 2017.

NOTE TO EDUCATORS

Visit www.focusreaders.com to find lesson plans, activities, links, and other resources related to this title.

INDEX

bridges, 9–10
building codes, 17–18, 20

doors, 28

floodgates, 29
foundations, 6, 19, 21

Galveston, Texas, 24

Hurricane Andrew, 20
hurricane clips, 19

Hurricane Katrina, 5–7, 10, 14
Hurricane Michael, 20

levees, 5–7, 10–12, 29

New Orleans, Louisiana, 5–7, 10–11, 14–15

paving materials, 13
pipe systems, 12–13, 14
power outages, 23–24

rain, 12, 14

satellites, 25
stilts, 20–21
storm surge, 5–7, 10–13, 24
storm windows, 18

wetlands, 28
winds, 5, 9, 12, 17–20, 27–28

Answer Key: 1. Answers will vary; 2. Answers will vary; 3. C; 4. A

32